YOUR KNOWLEDGE HAS VALUE

Amalia Aventurin

Archimedes Principle, Gaspycnometer and Geopycnometer

Petrophysics

GRIN Verlag

Bibliografische Information der Deutschen Nationalbibliothek:

Die Deutsche Bibliothek verzeichnet diese Publikation in der Deutschen National-
bibliografie; detaillierte bibliografische Daten sind im Internet über http://dnb.d-
nb.de/ abrufbar.

Imprint:

Copyright © 2013 GRIN Verlag GmbH
Druck und Bindung: Books on Demand GmbH, Norderstedt Germany
ISBN: 978-3-656-64489-7

This book at GRIN:

http://www.grin.com/en/e-book/272607/archimedes-principle-gaspycnometer-and-
geopycnometer

GRIN - Your knowledge has value

Der GRIN Verlag publiziert seit 1998 wissenschaftliche Arbeiten von Studenten, Hochschullehrern und anderen Akademikern als eBook und gedrucktes Buch. Die Verlagswebsite www.grin.com ist die ideale Plattform zur Veröffentlichung von Hausarbeiten, Abschlussarbeiten, wissenschaftlichen Aufsätzen, Dissertationen und Fachbüchern.

Visit us on the internet:

http://www.grin.com/

http://www.facebook.com/grincom

http://www.twitter.com/grin_com

Archimedes Principle, Gaspycnometer and Geopycnometer

I. Archimedes Principle measurement

The Archimedes method is used to determine the volume of an irregular shaped solid object. This is done by determining the dry mass of an object (which is given), the fully water saturated mass, measured with the Kern 572, and the mass of the sample when hanging in a water-filled bowl. Both measurements – the saturated and hanging-mass in a water-bowl – were done five times each by about 19.5°C air temperature and about 18°C water temperature. For this experiment we used two samples: G1 is a black stone with small mica particles and bigger white quartz inclusions. This stone is coarse-grained and compacted and therefore it could be a gabbro. G2 is a greenish sandstone with small particles and lesser compaction. The results of the measurements are shown in table 1.

	Saturated weights [g]	Immerged weights [g]
	1676.4	612.6
G1	1676.8	612.8
	1676.4	613.0
	1676.6	613.4
	1676.8	613.6
Average	1676.60	613.08
	Saturated weights [g]	**Immerged weights [g]**
G2	1985.5	777.4
	1985.6	778.4
	1985.8	778.6
	1986.0	779.2
	1985.4	779.4
Average	1985.66	778.60

Tab. 1: Measured masses by the Archimedes principle. The "saturated weights" is the mass of the sample by a water-saturation over one week. The "immerged weights" show the mass by hanging the sample in a water-filled bowl.

The Archimedes principle is based on the fact that an object immersed in a fluid will displace a certain amount of fluid. The amount of given fluid displaced depends on the density of the object. If the density of the fluid (ρ_{fluid}) and the mass of the object (m_{dry}) is known, you can write the formula like this:

$$\rho_{obj} = m_{dry}/V_b = m_{dry} / [(m_{sat} - m_{imm})/\rho_{fluid}] \tag{1}$$

Examples:

G1: ρ_{obj} = 1674.5 / 1063.6 = 1.574 g/cm^3

G2: ρ_{obj} = 1901.6 / 1210.17 = 1.571 g/cm^3

In order to determine the object's density (ρ_{obj}), the parameter to determine is the mass of the object immersed in the fluid (m_{imm}). In case of a rock sample with certain porosity you can use equation (1) to determine the bulk density (correlates to ρ_{obj}) or bulk volume of the sample. In order to calculate the matrix density (ρ_{ma}) the following equation can be written like this:

$$\rho_{ma} = m_{sat} / V_{ma} = (m_{sat} \cdot \rho_{fluid}) / [m_{dry} - (m_{sat} - m_{imm})] \tag{2}$$

Examples (with used ρ_{fluid}= 0.9982 g/cm^3):

G1: ρ_{ma} = (1676.4 * 0.9982) / [1674.5-(1676.4-612.8)] = 2.737 g/cm^3

G2: ρ_{ma} = (1985.4 * 0.9982) / [1901.6 - (1985.4-777.4)] = 2.853 g /cm^3

Where m_{sat} is the mass of a core sample saturated with a given fluid, V_{ma} is the matrix volume and m_{imm} the mass of displaced fluid. With these formulations you can write:

$$\emptyset= V_p /V_b = (V_b - V_{ma}) / V_b = (m_{sat} - m_{dry}) / m_{imm} \tag{3}$$

to determine the porosity.

Examples:

G1: $\emptyset$ = (1676.4 - 1674.5) / 612.8 = 0.0031= 0.35%

G2: $\emptyset$ = (1985.4-1901.6) / 777.4= 0.1077 = 10.79%

$$V_p = (m_{sat} - m_{dry}) / \rho_{fluid}$$
$$\tag{4}$$

Displaced water mass = $m_{sat} - m_{imm}/ \rho_{fluid}$ (density of water = 0.9982 g/cm³ at 20°C).

Examples:

G1: $V_p = m_{dw}$= (1676.4- 612.8)/0.9982 = 2.14 cm³

G2: $V_p= m_{dw}$= (1985.4-777.4)/0.9982 = 83.95 cm³

All results with the used formulas are shown in table 2.

3

The results and the solution method of the standard deviation calculation and the propagation of uncertainty are shown in the following.

1. The standard deviation

$$s = \sqrt{\frac{\sum_{i=1}^{n}(x_i - x_m)^2}{n-1}}$$
(5)

With x_i = measured value / x_m = mean value of the series / n = number of measurements

Example calculation (for dry weight of sample G1); Results, calculated as above:

Dry weight of the samples (with 5 values for each sample):

G1:   mean = 1676.64 g   std.dev. = 0.1497 g

G2:   mean = 1985.66 g   std.dev. = 0.2154 g

$$s = \sqrt{\frac{(1676.4g-1676.64g)^2+(1676.8g-1676.64g)^2+(1676.6g-1676.64g)^2+(1676.6g-1676.64g)^2+(1676.8g-1676.64g)^2}{n-1}}$$

Measured values of water-surrounded sample (with 5 values for each sample):

G1:   mean = 613.12 g   std.dev. = 0.325 g

G2:   mean = 778.6 g   std.dev. = 0.7043 g

2. Propagation of uncertainty, after Gauß:

$$\Delta G = \sqrt{\left(\frac{\partial G}{\partial x}\Delta x\right)^2 + \left(\frac{\partial G}{\partial x}\Delta Y\right)^2 + \frac{\partial G}{\partial x}\Delta z)^2 + \cdots}$$
(6)

With ΔG = most probable uncertainty / Δx; Δy; Δz = calculated std. dev. of the different measurement series (see results of step 1). In this case, the equation consists of two potential uncertainties, the dry weight and the value of the water-surrounded sample. This is the formula for the general part. The error calculation for the density of the rock sample (7), the density of the material (8), the porosity (9) and the volume of the saturated sample (10) are done by transformed versions of equation (6).

$$\Delta\rho_{obj}(m_{sat}, m_{im}) = \sqrt{\left(\rho_{fl} * m_{dry} * \left(-\frac{1}{m_{sat}^2 - m_{im}}\right) * \Delta m_{sat}\right)^2 + \left(\rho_{fl} * m_{dry} * \left(-\frac{1}{m_{sat} + m_{im}^2}\right) * \Delta m_{im}\right)^2}$$
(7)

The results are G1=0.5811 kg/m³ and G2=1.5426 kg/m³

$$\Delta\rho_{ma}\left(m_{sat},m_{im}\right)=\sqrt{\left(\left(-\frac{\rho_{fl}}{m_{dry}-m_{im}+m_{sat}^2}\right)*\Delta m_{sat}\right)^2+\left(\left(-\frac{\rho_{fl}*m_{sat}}{m_{dry}-m_{sat}+m_{im}^2}\right)*\Delta m_{im}\right)^2}$$

(8)

The results using equation (8) are G1=1.4557 kg/m³ and G2=2.6737 kg/m³.

$$\Delta\phi\left(m_{sat},m_{im}\right)=\sqrt{\left(\left(\frac{m_{dry}}{m_{im}}\right)*\Delta m_{sat}\right)^2+\left(\left(-\frac{m_{sat}-m_{dry}}{m_{im}^2}\right)*\Delta m_{im}\right)^2}$$

(9)

The results for G1=0.0029% and G2=0.0045% using equation (9).

$$\Delta V_p\left(m_{sat}\right)=\sqrt{\left(\left(\frac{m_{dry}}{\rho_{fl}}\right)*\Delta m_{sat}\right)^2}$$

(10)

The results using equation (10) are G1=2.5112·10⁻⁷ m³ and G2=4.1034·10⁻⁷m³.

5

	Bulk Volume [cm³]	Pore Volume [cm³]	Matrix Density [g/cm³]	Bulk Density [g/cm³]	Porosity [%]
G1	613.90	1.90	2.730	2.727	0.31
	613.90	2.30	2.741	2.727	0.37
	614.10	2.10	2.739	2.726	0.34
	614.50	2.10	2.737	2.724	0.34
	614.70	2.30	2.738	2.724	0.37
Average	**614.22**	**2.14**	**2.737**	**2.725**	**0.35**
G2	778.80	83.95	2.857	2.441	10.77
	779.80	84.15	2.854	2.438	10.79
	780.00	84.35	2.854	2.437	10.81
	780.60	84.55	2.853	2.436	10.83
	780.80	83.95	2.849	2.435	10.75
Average	**780.00**	**83.95**	**2.853**	**2.437**	**10.79**

Tab. 2: Calculated values with the Archimedes method. The used formulas are written at the bottom and where taken from Peters (2007) and Turcotte (2002). The used formulas are: For the bulk volume (v_b; $v_b = (m_{sat} - m_{imm})/\rho_{fl}$), for the pore volume ($v_p$; $v_p = (m_{sat} - m_{dry})/\rho_{fl}$), for the matrix density (ρ_m; $\rho_m = (m_{sat} \cdot \rho_{fluid})/[m_{dry} - (m_{sat} - m_{imm})]$), for the bulk density (ρ_{bulk} or ρ_{obj}; $\rho_b = m_{dry} / V_b$) and for the porosity (Φ; $\Phi = V_p / V_b$). Further values used for the calculation: G1 m_{dry} = 1674.5 g/cm³, G2 m_{dry} = 1901.6 g/cm³ and ρ_{water} = 0.9982 g/cm³ at 20°C.

From the low porosity and the quite high density values, we can conclude that sample G1 is either a highly compacted sedimentary rock like a dolomite (ρ=2.7-2.85 g/cm³ based on literature values) or a metamorphic rock, like a gabbro (ρ=2.6-2.85 g/cm³ based on literature values). But from the rock description, it is clearly that it's a gabbro. The bulk density of 2.437 g/cm³ (G2) points towards a rock of sedimentary origin. Turcotte and Gerald (2002) give values of 2.1-2.7 g/cm³ for shales and 1.9-2.5 g/cm³ for sandstones. The porosity (10.79%) also indicates that sample G2 is a sedimentary rock. A decision whether the sample is a quite compacted/cemented sandstone or a shale, cannot be made just from porosity and density measurements. Judging from the sample description the sample should be classified as a sandstone. For the error calculations, the standard deviation can be used. By using the measure values in the different equations, we achieve an error propagation which can be calculated by using the dependent case of the Gauß error propagation, which was already shown. We obtain partly very lower error values which show that we have also a very lower error propagation, i.e. the series of values we achieved in the first place are reliable and we have a small distribution of the values in a statistical point of view. So the Archimedes principle is a sure and simple method to tell something about porosity in a limited way. By further consideration the automatically measurements are more accurate, considering that the Archimedes method just involves the surface connected pores, whereas the geopycnometer and gaspycnometer methods also includes the non-connected pores.

II. Gaspycnometer measurement

The AccuPyc 1330 by Micromeritics measures automatically pressure changes of a solid and dried sample and determines the dry weight of the sample and their density. This is done by a displacement-reaction of the noble gas helium (He). For this experiment it's important that the samples are completely dry and free of volatiles. The sample with the unknown volume is placed in a chamber with a known volume. A second chamber with a known volume is filled with helium, which causes a higher pressure than the first one. The chambers are at the beginning disconnected from each other. After a valve opens, the gas diffuses into the first chamber and causes an increase in pressure, so the pressure in the second chamber must decrease. With the ideal gas law,

$$p \cdot V = n \cdot R \cdot T \tag{11}$$

where p is the pressure, V the volume, n is the number of gas atoms, R the Boltzmann constant and T is temperature in Klevin, the machine calculates the volume of the sample. The chambers are completely isolated, so there is no loss of gas molecules (n). The temperature is constant by about 19.5°C. Some deviations are possible; because there were about 10 people in the room while the measuring. For this experiment we used the two samples G1 and G2 from former measurements. The results are shown in table 3.

	G1	G2
Weight	24.681g	28.707001g
Volume	8.975528 cm³	10.323365 cm³
Bulk density	2.749815 g/cm³	2.780816 g/cm³
Number of measurements	5	5

Tab. 3: Results of the gaspycnometer measurement with AccuPyc 1330 for the samples G1 and G2.

For this reaction the inert gas helium is used, because of its small molecules, which diffuses fast in the even smallest pores. Other chemical elements which can be used are fluorine (F^-) or hydrogen (H^-), which also have small molecules. Due to their loading and their high reactivity they would react with the surface of the sample, this would lead to a chemical reaction which is not wished. Because of this it's better to use a noble gas, which is inert.

III. Geopycnometer measurement

The next step is a measurement by the GeoPyc 1360. This can just be done as a second step, because we need the results for the bulk density and the weight for the samples out of the Gaspycnometer. Otherwise this machine can't calculate its parameters: The bulk volume, the bulk density, the specific pore volume, porosity and the sample volume in the bulb. This apparatus "automatically determines bulk volume and density of a solid object" by a displacement-reaction "of a solid medium" (s. script). A bulb is filled with the graphite-like powder medium and put into the machine. As a first step, the machine compacts the bulb without a sample by rotating and pushing with a piston. The bulb has a diameter of 50 mm and the machine compresses the powder with strength of 135 N and 2.0387 cm²/nm. After compaction the apparatus measures the named parameter of the powder for calibration. The calibration has to be done for every measurement. Then the bulb is filled with one of the samples. The program of the machine runs a second time and measures now the parameter for the sample. Before you start to measure you have to be aware, that the sample is fully covered by the powder – otherwise the piston get crushed – and that you don't have a loss of the powder with which you have done the calibration – otherwise the measurement fails. The results of this measurement are shown in table 4. The temperature is here also constant by about 19.5°C. Some deviations are possible, because there were about 10 people in the room while the measuring.

	G1	G2
Average bulk volume	8.7707 cm³	11.7256 cm³
Average bulk density	2.8140 g/cm³	2.4482 g/cm³
Specific pore volume	-0.0083 cm³/g	0.0488 cm³/g
Porosity	-0.0083 %	11.959 %
Sample volume in the bulb	8.439 %	10.982 %

Tab. 4: measurement-results of the geopycnometer GeoPyc 1360 for both samples G1 and G2.

The results for the volume and the bulk density are for both samples with both methods nearly the same. There are no bigger differences to complain about. The results of the Geopycnometer for porosity and specific pore volume for sample G1 are negative, which just indicates that there is no porosity. This result approves the theory of a highly compacted gabbro. Even the density goes together with density of gabbro from the literature (2800-3000 kg/m³ / 2.50-3.30 g/cm³). For the sample G2 the density goes together with densities from the literature (2.00-2.80 g/cm³), but the porosity is too high for a "normal" sandstone (30-40%). Even here we have a higher compaction in maybe higher depths, than we first thought. The low specific pore volume is an indicator for this, too.

Other methods to determine the density are with the aerometer or with a pycnometer. Both methods base on the Archimedes-principle: Weighing first a water-filled bowl and then the dipping glass-bowl with the sample in it.

Some methods to determine the porosity are the "apparent density method, water pycnometry, Hg porosimetry and the gas penetration technique" (Palacio et al. 1999). There exist several and different methods to determine the porosity, but all base on the same equation:

$$\theta \, [\%] = 1 - \frac{\rho}{\rho_0} \bullet 100\% \tag{12}$$

where ρ is the bulk density and ρ_0 is the particle density.

IV. Conclusion

The Archimedes method is good to determine the saturated and immerged weight of a sample, but this takes a long time. For a quick determination it is better to use the Geo- and Gaspycnometer methods and with the apparatus you have a lower error margin. But for every of the three methods you have first to determine the dry weight of the samples, what takes a lot of time. The automatic pycnometer methods allow a determination of bulk density, matrix density and porosity, which the Archimedes' method is not able to.

But with the focus on the question, which results are more reliable it can be said, that the answer is the pycnometer-measurement. With the Archimedes-method you have too many sources of error, e.g. the loss of water by the "transport" of the sample to the scales and out of the water-bowl again. Even the toweling of the sample before you transport them causes a loss of water and a drying of the sample. The movement of the water causes also inaccuracy of the not 100% correct scales, which fluctuates also because of the movement of the stone, which is just hanged in the water. Even the pycnometers have some inaccurateness, mainly caused by variations in temperature in the room and by a failed calibration, but besides of this the pycnometers work with fewer errors.

In comparison the pycnometric methods are able to measure the whole porosity respectively the gas through-flow-able pores. With the Archimedes method only the water through-flow can be determined. All three measurement methods are good to calculate the porosity and the density of a sample.

Never the less which method you use depends on the questioning, if you want to determine porosity and bulk density, etc. and which source you want to use, like water or a gas. There exist several methods and apparatus', e.g. the "boost"-method, different qualitative and quantitative methods, which mostly base on the same principles that we used. To describe the several methods in detail would excess the report.

Our results seem to be realistic in comparison to literature values. The calculated values are a little bit higher than the reference values but not with stronger deviations.

V. References

- http://www.seilnacht.com/versuche/dichteb.html (12.12.2012)
- http://www.geo-glossar.de/woerterbuch/porositaet.html (12.12.2012)
- http://www.geo.fu-berlin.de/fb/e-learning/petrograph/tabellen/gesteinsdichte.html (12.12.2012)
- http://www.stone-park.de/fileadmin/pdf/Gesteinskunde.pdf (14.12.2012)
- L. Palacio, P. Prádanos, J.I. Calvo, A. Hernández, 1999. Porosity measurements by a gas penetration method and other techniques applied to membrane characterization. Thin Solid Films Vol. 348, 22–29.
- Peters, E. J. (2007). Petrophysics. University of Texas.
- Schubert, G., Turcotte, D. L., 2002. Geodynamics. Cambridge University Press. second edition.